A Student Guide

to

Man's Selection:

Charles Darwin's Theory

of

Creation, Evolution, and Intelligent Design

by

Marc Watson and Barbara Angle

A Student Guide to the larger companion work *Man's Selection: Charles Darwin's Theory of Creation, Evolution, and Intelligent Design,* by Marc Watson

Table of Contents

PREFACE
Creation, Evolution, and Intelligent Design
Two Sides of the Question

> *... but I look with confidence to the future, to young and rising naturalists, who will be able to view both sides of the question with impartiality.*
>
> Darwin, Charles. The origin of Species, Sixth Edition (1872) page 423

What do you think regarding fundamental questions about life and the purpose for human existence? Who or what created life? Why are we here? How are we supposed to behave? Is there a plan of creation? For centuries questions regarding creation and evolution have spawned a variety of theories by scientists, philosophers, and theologians. Driven by strong beliefs, proponents of alternative theories have segregated into two highly partisan schools of thought regarding questions on the fundamental causation for both creation and evolution. The two sides of this question may be referred to as the **"Case for Chance,"** and the **"Case for Design."** This inquiry does not ask you to regurgitate what you've been told. Instead, it asks you to be a SMART learner - to use Science, Math, Art, Reason, and Tradition to answer the question - What do you think?

Why does this matter? For thousands of years, questions regarding creation and evolution have shaped human beliefs and driven human behavior regarding core concepts of good and evil; right and wrong. Such questions are not esoteric. Scientists, philosophers, and theologians have debated these questions, some claiming that they alone can provide the answers. History, however, has demonstrated that no single discipline - science, philosophy, or theology - has a monopoly on truth regarding the creation of the universe or the phenomenon we call life. This inquiry is fundamental to our understanding of earth's immensely diverse and highly integrated biosphere.

More importantly, this inquiry must consider whether or not the modern human species is a meaningless form of ape whose superior intellect is a product of our struggle for existence. Was our evolution driven exclusively by the Survival of the Fittest? Did our unique level of intelligence evolve so that we can fiercely compete with and dominate our competition, even if it brings our world to the brink of destruction? Or are modern humans purposeful creations endowed with a unique intellect - a Singular Intelligence? Has this Singular Intelligence prompted us to establish laws and convey rights that transcend nature's laws and the singular right of survival? If so, might we use this gift to advance our understanding of Intelligent Design, positively shape the natural world for the benefit of all who dwell on our planet, and discern what Charles Darwin called the *"plan of creation"*? If the 20th Century taught us anything, it is that these are pragmatic questions. How you answer these questions can have a major impact on the future of humanity.

How is this book different? This Student Guide provides a scientific inquiry into Charles Darwin's phenomenal observations on Creation, Evolution, and Intelligent Design. This is a scientific inquiry, and therefore it does not present philosophical or theological arguments; not that such arguments are not influential on our thinking. Accordingly, we must recognize that individual beliefs may impact the objectivity of scientific observation and reasoning. Science is an intellectual journey in which we are challenged to consider alternatives, weigh evidence, and defend our position in the quest for knowledge. Charles Darwin was confident that future scientists would gain a greater understanding of creation and evolution by viewing both sides of the question with impartiality.

On one side of this question you may find those who assert that the original creation of life, and the evolution of every life form that exists or has ever existed is the result of an **U**nintentional, **R**andom, **A**ccidental, **H**appenstance without purpose or meaning (we will refer to this theory as **"URAH"**). Those taking this position often insist that Natural Selection is the singular means of evolutionary modification and has been the exclusive means of the creation of all living things. This camp is sometimes referred to as evolutionists, or "Darwinists".

On the other side of the question you may find those who claim that the original creation of life, and the creation of every life form that exists or has ever existed is the result of **Individual Divine Acts** of a Supernatural Being, with each and every act being part of a predetermined plan. Those taking this position often insist that species are not mutable, Natural Selection does not occur, and new species cannot evolve from progenitor species. This camp is sometimes referred to as "creationists."

Darwin did not subscribe to either extreme. His findings were based on rigorous scientific observation and therefore can serve as the basis for inquiry into this largely unknown area of science. The foundation for Darwin's study was a process he described as *"man's power of accumulated selection,"* referred to herein simply as **"Man's Selection"** (Note: Darwin's use of the term *"man"* in this context was not intended to be gender specific, but was instead a genderless reference to humanity.) Through the study of Man's Selection, he observed that multiple forms of **Variation** were the catalyst for creation. He also observed that several types of **Selection** preserved the most favorable variations for adaptation by the organism, preserving traits that could be inherited by its offspring and passed along to future generations. Darwin's theory supposes that complex combinations of Variation and Selection, governed by *"laws impressed on matter by the Creator,"* have resulted in the creation and evolution of life.

Darwin's work leads us to a third theory that fits between the extremes. Under this theory, the creation and evolution of life are the result of a series of integrated processes that make up a **R**ational, **U**niform, **A**ccurate, and **H**olistic system (we will refer to this theory as **"RUAH".)** Darwin bridged the "Case for Chance" with the "Case for Design" through his analogy of an architect who masterfully creates noble and purposeful edifices from seemingly meaningless uncut stones produced by natural laws unrelated to the edifice created (see Darwin's observation on page 34 of this guide.)

This inquiry takes a fresh look at Darwin's famous trilogy on evolution, using the latest editions of the following three works:

(1) *The Origin of Species* (originally *On The Origin of Species*). 1872;
(2) *The Variation of Animals and Plants Under Domestication*, 1875; and,
(3) *The Descent of Man and Selection Related to Sex 1874*.

These publications, along with selected quotes from Thomas Huxley, referred to as Darwin's bulldog, and Alfred Russel Wallace, who co-discovered the Theory of Natural Selection with Darwin, will round out this inquiry. Darwin was convinced that the study of Man's Selection was of great value. Evaluating Darwin's work anew allows us to consider both sides of the question and advance our understanding of life and the human condition.

Why should you care? You are the future. You are one of the *"young and rising"* scientists that Charles Darwin was confident could view both sides of the evolution question with impartiality. You are also a **SMART** learner - endowed with Singular Intelligence. You can use Science, Math, Art, Reason, and Tradition to advance your understanding of the complex issues that shape our world. Was life spawned as a happenstance of matter and natural law, or was it *"originally breathed by the Creator"* into one or more forms, as Charles Darwin observed? Are humans simply a twig on the evolutionary tree - a meaningless form of ape? Or were we created for a higher purpose through a process analogous to **Man's Selection.** The answers to such questions compel our consideration of moral and ethical behavior as the foundation of a rule of law and human rights that transcend the singular law of nature and the natural right of the Survival of the Fittest - the driving force of **Natural Selection**.

You will be required to apply **S**cientific observation, **M**athematical logic, communicative **A**rt, critical **R**easoning, and the wisdom of thousands of years of **T**radition to objectively analyze one of the most controversial and hotly debated topics of all times: The Creation and Evolution of life. Will you allow the answers to such questions to be dictated to you? Or, will you be SMART enough to consider impartially both sides of the question and determine for yourself where the evidence falls with regard to the Case for Chance or the Case for Design. Lead or follow - it's your choice.

A Matter For Science. Some question whether this is a matter for scientific inquiry. In 1972, the California School Board asked America's top rocket scientist, Wernher Von Braun, the former head of NASA's Marshall Space Flight Center and recipient of the National Medal of Science, for his view on this matter. The following is an excerpt from his reply.

In response to your inquiry about my personal views concerning the "Case for DESIGN" as a viable scientific theory for the origin of the universe, life, and man, I am pleased to make the following observations.

For me, the idea of a creation is not conceivable without evoking the necessity of design. ...

... To be forced to believe only one conclusion - that everything in the universe happened by chance - would violate the very objectivity of science itself.

...

I have discussed the aspect of a Designer at some length because it might be that the primary resistance to acknowledging the "Case for Design" as a viable scientific alternative to the current "Case for Chance" lies in the inconceivability, in some scientists' minds, of a Designer. The inconceivability of some ultimate issues (which will always lie outside scientific resolution) should not be allowed to rule out any theory that explains the interrelationship of observed data and is useful for prediction.

We in NASA were often asked what the real reason was for the amazing string of successes we had with our Apollo flights to the Moon. I think the only honest answer we could give was that we tried to never overlook anything. It is in that same sense of scientific honesty that I endorse the presentation of alternative theories for the origin of the universe, life, and man in the science classroom. It would be an error to overlook the possibility that the universe was planned rather than happened by chance.

INTRODUCTION
On Darwin's Work

Student Inquiry: Should all of Charles Darwin's work be considered in the scientific study of evolution and the Theory of Natural Selection?

Before you start this inquiry, consider how your beliefs or prior knowledge may influence your scientific observations or your understanding of the world around you. Write down your thoughts to the following questions as you prepare for this inquiry.

Intro.1: Are you familiar with the Case for Chance and the Case for Design? If so, how would you describe each?

Intro.2: Consider what you have been taught thus far about evolution science. Do these teachings align with your views of the creation of life?

Intro.3: What knowledge do you have of Charles Darwin and his work on evolution theory?

Charles Darwin began his seminal work on Natural Selection with the following quotations from highly influential thinkers of his time.

> *"But with regard to the material world, we can at least go so far as this - we can perceive that events are brought about not by insulated interpositions of Divine power, exerted in each particular case, but by the establishment of general laws."*
>
> Whewell: *Bridgewater Treatise*
>
> *"The only distinct meaning of the word 'natural' is stated, fixed, or settled; since what is natural as much requires and presupposes an intelligent agent to render it so, ie, to effect it continually or at stated times, as what is supernatural or miraculous does to effect it for once."*
>
> Butler: *Analogy of Revealed Religion*
>
> *"To conclude, therefore, let no man out of a weak conceit of sobriety, or an ill-applied moderation, think or maintain, that a man can search too far or be too well studied in the book of God's word, or in the book of God's works; divinity or philosophy; but rather let men endeavour an endless progress or proficience in both."*
>
> Bacon: *Advancement of Learning*
>
> Darwin, Charles. *The Origin of Species*, Introductory Quotations

As you go through this inquiry, you may find that conducting just a little research on several historical characters, including the three men above, may add significantly to your understanding of the subject. Such research may also help you support or defend your position in discussions on creation, evolution, and intelligent design.

Intro 4: Using a search engine, conduct research on the three men who are credited with the above quotes. Then, on the notes pages at the end of this chapter, try to paraphrase or restate in your own words each of the above quotations. Use the notes pages at the end of each chapter to collect all of your notes or research as you conduct this inquiry.

Intro.5: Why do you think Charles Darwin included the above three passages at the very beginning of *"The Origin of Species"*?

Intro.6: What do you think is meant by the following phrases: *"insulated interpositions of Divine power," "the establishment of general laws,"* and *"an intelligent agent?"*

Intro.7: How does Joseph Butler compare natural and supernatural occurrences?

Intro.8: How might Bacon's cautionary statement impact an inquiry into this controversial subject?

Intro.9: Does Darwin's inclusion of these three passages in *The Origin of Species* have any impact on your understanding of his observations regarding the creation and evolution of life?

The following two passages are taken from the *"Recapitulation and Conclusion"* of *The Origin of Species.*

There is grandeur in this view of life, with its several powers, having been originally breathed by the Creator into a few forms or into one; and that, whilst this planet has gone cycling on according to the fixed law of gravity, from so simple a beginning endless forms most beautiful and wonderful have been, and are being evolved.

Darwin, Charles. *The Origin of Species*, Page 428 - 429.

I have now recapitulated the facts and considerations which have thoroughly convinced me that species have been modified, during a long course of descent. This has been chiefly through the natural selection of numerous successive, slight, favourable variations; aided in an important manner by the inherited effects of the use and disuse of parts; and in an unimportant manner, that is in relation to adaptive structures, whether past or present, by the direct action of external conditions, and by variations which seem in our ignorance to arise spontaneously. It appears that I formerly underrated the frequency and value of these latter forms of variation, as leading to permanent modifications independently of natural selection. But as my conclusions have lately been much misrepresented, and it has been stated that I attribute the modification of species exclusively to natural selection, I may be permitted to remark that in the first edition of this work, and subsequently, I placed in a most conspicuous position - namely, at the close of the introduction - the following words: "I am convinced that natural selection has been the main but not the exclusive means of modification". This has been to no avail. Great is the power of steady misrepresentation; but the history of science shows that fortunately this power does not long endure.

Darwin, Charles. *The Origin of Species*, page 421

Consider the quotations Darwin used at the beginning of his work, and the passages from the last section of his work shown on the previous page. Now consider that Joseph Butler's quotation was not included in the first edition of *On The Origin of Species*, but was added by Darwin to the 2nd and every subsequent edition.

Intro.10: Why do you think Darwin added Butler's quote to *The Origin of Species*?

As with the addition of Joseph Butler's quote, Darwin added the term *"the Creator"* to the final passage of the 2nd and subsequent editions of *The Origin of Species*, as shown on the previous page.

Intro.11: Why do you think Darwin added the term *"the Creator"* to the concluding passage of *The Origin of Species*?

Intro.12: In Darwin's view, what are the means through which species have been modified?

Intro.13: What forms of variation *"leading to permanent modifications independently of natural selection,"* did Darwin feel he had underrated with regard to frequency and value?

Intro.14: According to Darwin, how did life, with its several powers, originate; and through how many forms was life originally created?

Intro.15: How did Darwin feel his conclusions had been misrepresented? What did he do as a response to such misrepresentation?

Intro.16: In Darwin's opinion, what has the history of science shown?

Student Inquiry: Should all of Charles Darwin's work be considered in the scientific study of evolution and the Theory of Natural Selection?

NOTE: Several of Darwin's passages are used multiple times throughout this Guide. Such passages are replicated so that you might consider different aspects of the passage, or view the passage with new knowledge that you have acquired in your journey through this inquiry.

Notes Pages are included between each chapter. Consider recording your thoughts and research on this subject as you go through each chapter of this inquiry.

CHAPTER 1
Darwin on Man's Selection

Student Inquiry: What powerful principle did Charles Darwin claim was the key to understanding his theory of Natural Selection?

> *One of the most remarkable features in our domesticated races is that we see in them adaption, not indeed to the animal's or plant's own good, but to man's use or fancy. ... The key is man's power of accumulated selection: nature gives successive variations; man adds them up in certain directions useful to him. In this sense he may be said to have made for himself useful breeds.*
>
> *The great power of this principle of selection is not hypothetical.*
>
> Darwin, Charles. *The Origin of Species*, page 22
>
> *We have seen that man by selection can certainly produce great results, and can adapt organic beings to his own uses, through the accumulation of slight but useful variations, given to him by the hand of Nature.*
>
> Darwin, Charles. *The Origin of Species*, page 49.

This book uses the term **"Man's Selection"** as the shortened form of what Darwin observed as *"man's power of accumulated selection."*

1.1: What did Darwin observe is the key to the creation of remarkable features in our domestic species?

1.2: For what purpose are domestic species created?

> *It is therefore, of the highest importance to gain a clear insight into the means of modification and coadaptation. At the commencement of my observations it seemed to me probable that a careful study of domesticated animals and of cultivated plants would offer the best chance of making out this obscure problem. Nor have I been disappointed; in this and in all other perplexing cases I have invariably found that our knowledge, imperfect though it be, of variation under domestication, afforded the best and safest clue. I may venture to express my conviction of the high value of such studies, although they have been very commonly neglected by naturalists.*
>
> *From these considerations I shall devote the first chapter of this Abstract to Variation under Domestication.*
>
> Darwin, Charles. *The Origin of Species*, page 3.

1.3: What types of species (or races) are created through Man's Selection?

1.4: What did Darwin propose offered the best chance to *"gain a clear insight into the means of modification and coadaptation?"*

1.5: What process is used to create domesticated animals and cultivated plants?

> *If selection consisted merely in separating some very distinct variety, and breeding from it, the principle would be so obvious as hardly to be worth notice; but its importance consists in the great effect produced by accumulation in one direction, during successive generations, of differences absolutely inappreciable by an uneducated eye - differences which I for one have vainly attempted to appreciate. Not one man in a thousand has accuracy of eye and judgment sufficient to become an eminent breeder. If gifted with these qualities, and he studies his subject for years, and devotes his lifetime to it with indomitable perseverance, he will succeed, and may make great improvements; if he wants any of these qualities he will assuredly fail.*
>
> Darwin, Charles. *The Origin of Species,* page 23.

1.6: According to Darwin, can Man's Selection be achieved by simply separating and breeding distinct varieties?

1.7: What does it take to succeed at employing the process of Man's Selection and become an eminent breeder?

1.8: Based on Darwin's observations, does the process of Man's Selection and the creation of a domestic species require an intelligent agent?

> *I have called this principle, by which each slight variation, if useful, is preserved, by the term Natural Selection, in order to mark its relation to man's power of selection. But the expression often used by Mr. Herbert Spencer, of the Survival of the Fittest, is more accurate, and is sometimes equally convenient. We have seen that man by selection can certainly produce great results, and can adapt organic beings to his own uses, through the accumulation of slight but useful variations, given to him by the hand of Nature. But Natural Selection, we shall hereafter see, is a power incessantly ready for action, and is as immeasurably superior to man's feeble efforts, as the works of Nature are to those of Art.*
>
> Darwin, Charles. *The Origin of Species,* page 49

1.9: Why did Darwin call the principle by which each slight variation, if useful, is preserved, by the term Natural Selection?

> *Can the principle of selection, which we have seen is so potent in the hands of man, apply under nature? I think we shall see that it can act most efficiently.*
>
> Darwin Charles, *The Origin of Species*, Sixth Edition (1876), page 62

1.10: Did Darwin believe that nature could apply the same principle of selection as Man's Selection?

1.11: Compare and contrast the processes of Man's Selection and Natural Selection using the System Digram Below.

1.12: Compare the primary purpose for retention of traits in domestic species versus the primary purpose for retention of traits in wild species.

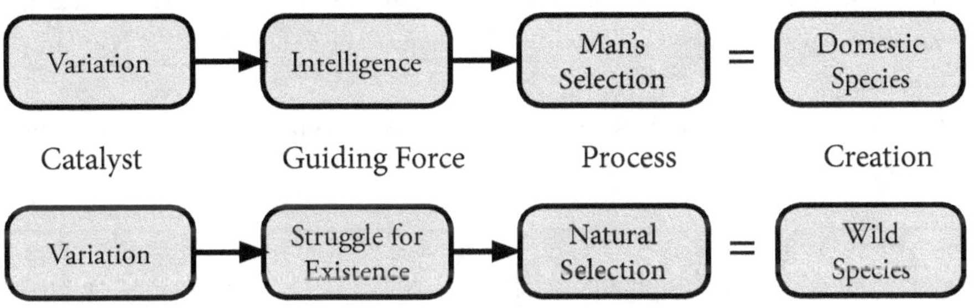

> *Can we wonder, then, that Nature's productions should be far "truer" in character than man's productions; that they should be infinitely better adapted to the most complex conditions of life, and should plainely bear the stamp of far higher workmanship?*
>
> Darwin, Charles. *The Origin of Species*, page 65.

1.13: How would you define the term *"workmanship"*?

1.14: What does Darwin's use of the term workmanship suggest with respect to any skill required in the creation of Nature's productions?

1.15: How does Darwin compare Nature's productions with man's productions?

Answering the Inquiry: What powerful principle did Charles Darwin claim was the key to understanding his theory of Natural Selection?

Creation and Evolution: A System of Integrated Processes

Creation and evolution are both the result of the same system of integrated processes. The diagram on the previous page shows the elements common to all. As you go through this guide, you may want to refer back to this page and make notes regarding how elements of each process fit together in an integrated system to create or evolve new species.

Elements of the Creation and Evolution System Diagram

1. Variation Processes: Consider both the causes and changes of each different type of variation - every variation must have a cause.

2. Guiding Force: There must be a guiding force that determines which type of Selection will act on any variation. These forces range from the struggle to survive, to an intelligent agent, to the laws impressed on matter by the Creator, such as the law of correlated growth.

3. Selection Processes: Which type of Selection has preserved the characteristics of the variant or species, and what allows the Selections to be accumulated in a certain direction.

4. Creation: At what point does an entirely new species come into existence - in other words, when is a new species Created?

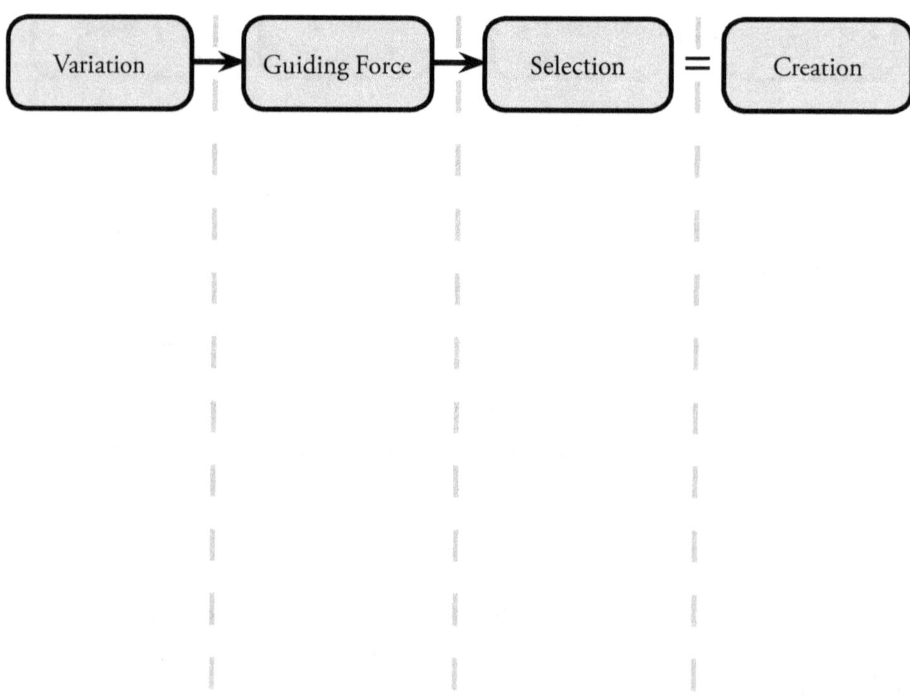

Notes:

CHAPTER 2
Darwin on Variation

Student Inquiry: What is the catalyst for the Creation and Evolution of all life?

> *The power of Selection, whether exercised by man, or brought into play under nature through the struggle for existence and the consequent survival of the fittest, absolutely depends on the variability of organic beings. Without variability nothing can be effected;...*
>
> Darwin, Charles. *Variation Under Domestication,* Volume II, Page 176.
>
> *But when man is the selecting agent, we clearly see that the two elements of change are distinct; variability is in some manner excited, but it is the will of man which accumulates the variations in certain directions; and it is this latter agency which answers to the survival of the fittest under nature.*
>
> Darwin Charles. *The Origin of Species,* Pages 107-108

2.1: According to Darwin, what is required before the power of Selection, whether exercised through Man's Selection or Natural Selection, can be effected?

2.2: Through Man's Selection, Darwin observed two distinct elements of change in the creation and evolution of species. Name those elements.

2.3: Which element comes first and why?

2.4: Consider the process diagram on page 15. How are Natural Selection and Man's Selection alike and how are they different?

> *I have hitherto sometimes spoken as if the variations - so common and multiform with organic beings under domestication, and in a lesser degree with those under nature - were due to chance. This of course, is a wholly incorrect expression, but it serves to acknowledge plainly our ignorance of the cause of each particular variation.*
>
> Darwin, Charles. *The Origin of Species,* page106

2.5: What did Darwin observe to be a *"wholly incorrect expression"* regarding Variation?

> *Variability is governed by many complex laws, - by correlated growth, compensation, the increased use and disuse of parts and the definite action of the surrounding conditions. There is much difficulty in ascertaining how largely our domestic productions have been modified; but we may safely infer that the amount has been large, and that modifications can be inherited for long periods.*
>
> Darwin, Charles. *The Origin of Species*, page. 410.

2.6: What governs Variability?

2.7: What two things may we safely infer about variation and the resulting modification of plants and animals?

Types of Variation

The first two chapters of *The Origin of Species* focused on **Variation**, and observations regarding variation are prevalent throughout Darwin's work. Darwin recognized that Variability is governed by many complex laws. He tackled this complicated process by breaking variation into several categories, including: (1) Individual Variations; (2) Variations caused by the use and disuse of parts or the actions of surrounding conditions; (3) Correlated Variations; and (4) Spontaneous Variations.

> *Individual Differences*
>
> *The many slight differences which appear in the offspring from the same parents, or which it may be presumed have thus arisen, from being observed in the individuals of the same species inhabiting the same confined locality, may be called individual differences. No one supposes that all the individuals of the same species are cast in the same actual mould. These individual differences are of the highest importance for us, for they are often inherited, as must be familiar to every one; and they thus afford materials for natural selection to act on and accumulate, in the same manner as man accumulates in any given direction individual differences in his domestic productions.*
>
> Darwin, Charles. *The Origin of Species*, page 34.
>
> *Under the term "variations," it must never be forgotten that mere individual differences are included. As man can produce a great result with his domestic animals and plants by adding up in any given direction individual differences, so could natural selection, but far more easily, from having incomparably longer time for action.*
>
> Darwin, Charles. *The Origin of Species*, page 64

> *There is one point connected with individual differences which is extremely perplexing: I refer to those genera which have been called "protean" or "polymorphic," in which species present an inordinate amount of variation.... These facts are very perplexing, for they seem to show that this kind of variability is independent of the conditions of life. I am inclined to suspect that we see, at least in some of these polymorphic genera, variations which are of no service or disservice to the species, and which consequently have not been seized on and rendered definite by natural selection, as hereafter to be explained.*
>
> Darwin, Charles. *The Origin of Species*, page 35

Individual variations are those that cause each of us to be an individual creation. Individual differences are those small changes in our DNA that make us slightly different from our parents and siblings.

2.8: How do humans produce a great result in our domestic animals from individual variations?

2.9: What type of individual differences did Darwin believe could not be acted upon or rendered definite by natural selection? Why not?

> *Changed habits produce an inherited effect, as in the period of flowering plants when transported from one climate to another. With animals the increased use or disuse of parts has had a more marked influence; thus I find in the domestic duck that the bones of the wing weigh less and the bones of the leg more, in proportion to the whole skeleton, than do the same bones in the wild-duck; and this change may be safely attributed to the domestic duck flying much less, and walking much more, than its wild parents.*
>
> Darwin, Charles. *The Origin of Species*, page 8
>
> *But we learn from the study of our domestic productions that the disuse of parts leads to their reduced size; and that the result is inherited.*
>
> Darwin, Charles. *The Origin of Species*, page 400

Darwin observed that environmental conditions and changed habits can cause variations in individual organisms that can be inherited by offspring. This process is known as the Inheritance of Acquired Characteristics (**"IAC"**), or Lamarckism.

2.10: Some scientists today do not accept IAC. Research IAC and Lamarckism. Then investigate the "Giraffe Neck Mystery." Do you think that an animal's evolution can in part be driven by the animal's behavior?

Correlated Variation

I mean by this expression that the whole organisation is so tied together, during its growth and development, that when slight variations in any one part occur and are accumulated through natural selection, other parts become modified. This is a very important subject, most imperfectly understood, and no doubt wholly different classes of facts may be here easily confounded together.

Darwin, Charles. *The Origin of Species*, page 114

Even when selection has been applied by man to some one character alone -- of which our cultivated plants offer the best instances -- it will invariably be found that although this one part, whether it be the flower, fruit, or leaves, has been greatly changed, almost all the other parts have been slightly modified. This may be attributed partly to the principle of correlated growth, and partly to so-called spontaneous variation.

Darwin, Charles. *The Origin of Species*, page 170

With varieties and species, correlated variation seems to have played an important part, so that when one part has been modified other parts have been necessarily modified.

Darwin, Charles. *The Origin of Species*, page 415

2.11: What did Darwin observe regarding our understanding of correlated variation?

Under domestication we see much variability, caused, or at least excited, by changed conditions of life; but often in so obscure a manner, that we are tempted to consider the variations as spontaneous.

Darwin, Charles. *The Origin of Species*, page 410

In the earlier editions of this work I underrated, as it now seems probable, the frequency and importance of modifications due to spontaneous variability.

Darwin, Charles. *The Origin of Species*, page 171

2.12: How did Darwin admittedly underrate Spontaneous Variation?

2.13: If variation is not "by chance," (see quote on page 19), then what force might cause Spontaneous Variations?

2.14: How do multiple types of Variation contribute to the diversity of life?

Answering the Inquiry: What is the catalyst for the Creation and Evolution of all life?

Notes:

CHAPTER 3
Darwin on Creation and Selection

Student Inquiry: How do Darwin's observations address the Creation and Evolution of all living things?

Creation: The terms create and creation may often be controversial when used in the scientific study of life and evolution. For this work, we will use the definition of "Create" offered by Merriam-webster.com

transitive verb - 1: to bring into existence. 2.b: to produce or bring about by a course of action or behavior. 3: CAUSE. "

intransitive verb - 1: to make or bring into existence something new.

Based on this definition, everything new or different was at some point created, or brought into existence. The Universe, atoms, life, and all living things were at some point, created. Each of us are individual creations. The question for science is to try to understand the driving force or cause of the processes of creation. Darwin did not shy away from this difficult subject and addressed the question in *The Origin of Species*.

> *Authors of the highest eminence seem to be fully satisfied with the view that each species has been independently created. To my mind it accords better with what we know of the laws impressed on matter by the Creator, that the production and extinction of the past and present inhabitants of the world should have been due to secondary causes, like those determining the birth and death of the individual. ...*
>
> *There is grandeur in this view of life, with its several powers, having been originally breathed by the Creator into a few forms or into one; and that, whilst this planet has gone cycling on according to the fixed law of gravity, from so simple a beginning endless forms most beautiful and wonderful have been, and are being evolved.*
>
> Darwin, Charles. *The Origin of Species*, Page 428 - 429.

Based on Darwin's observations, consider the following:

3.1: What force impressed the laws on matter?

3.2: What force created the original form or forms of life?

3.3: What governs or guides secondary causes in the creation of life?

Darwin's observations were a matter for science. His conclusions on evolution were based on scientific reasoning instead of emotional feelings.

> *Another source of conviction in the existence of God, connected with the reason and not with the feelings, impresses me as having much more weight. This follows from the extreme difficulty or rather impossibility of conceiving this immense and wonderful universe, including man with his capacity of looking far backwards and far into futurity, as the result of blind chance or necessity. When thus reflecting I feel compelled to look at a First Cause having an intelligent mind in some degree analogous to that of man; and I deserve to be called a Theist.*
>
> *This conclusion was strong in my mind about the time, as far as I can remember, when I wrote the Origin of Species; and it is since that time that is has very gradually with many fluctuations become weaker. ...*
>
> *I cannot pretend to throw the least light on such abstruse problems. The mystery of the beginning of all things is insoluble by us; and I for one must be content to remain an Agnostic.*
>
> Darwin, Charles. *The Autobiography of Charles Darwin 1809-1882*, With original omissions restored, Edited with Appendix and Notes by his grand-daughter Nora Barlow; Collins, St. James Place, London (1958) pages 92 - 93.

3.4: What was Darwin's observation regarding the initiating force in the creation of the Universe and life when he wrote *The Origin of Species*?

3.5: Darwin found it extremely difficult if not impossible to conceive that this immense and wonderful universe, including humans, with their ability to look to both the past and the future, was a result of what?

3.6: Based on Darwin's writing, to which scientific theory do you think he subscribed: the Case for Chance or the Case for Design?

> *Therefore I cannot doubt that the theory of descent with modification embraces all the members of the same great class or kingdom. I believe that animals are descended from at most only four or five progenitors, and plants from an equal or lesser number.*
>
> *Analogy would lead me one step further, namely, to the belief that all animals and plants are descended from some one prototype. But analogy may be a deceitful guide. Nevertheless all living things have much in common, in their chemical composition, their cellular structure, their laws of growth, and their liability to injurious influences.*
>
> Darwin, Charles. *The Origin of Species*, page 424.

Darwin's observations regarding selection and evolution are based on the *"theory of descent with modification."* Under this theory, new species can be created by the introduction of **Variation** in existing species, and the **Selection** of the specific variations that will be inherited by and therefore preserved in future generations. Darwin further observed that this action *"embraces all the members of the same great class or kingdom."*

3.7: Based on Darwin's observations, animals have descended from at most ___ progenitor forms, and plants from ______________ number.

3.8: Based on Darwin's observations, do you think plant species and animals species may have been created from separate progenitors?

3.9: What caution did Darwin make regarding the analogous belief that all animals and plants are descended from some one prototype?

> *The principle of selection may be conveniently divided into three kinds. Methodical selection is that which guides a man who systematically endeavors to modify a breed according to some predetermined standard. Unconscious selection is that which follows from men naturally preserving the most valued and destroying the less valued individuals, without any thought of altering the breed; and undoubtedly this process slowly works great changes. Unconscious selection graduates into methodical, and only extreme cases can be distinctly separated; for he who preserves a useful or perfect animal will generally breed from it with the hope of getting offspring of the same character; but as long as he has not a predetermined purpose to improve the breed, he may be said to be selecting unconsciously. Lastly, we have Natural selection, which implies that the individuals which are best fitted for the complex, and in the course of ages changing conditions to which they are exposed, generally survive and procreate their kind. With domestic productions, natural selection comes to a certain extent into action, independently of, and even in opposition to, the will of man.*
>
> Darwin, Charles. *Variation Under Domestication,* Volume II Pages 177 - 178.

3.10: Darwin divided the principle of selection into three types and defined each. Name and briefly describe these three types of selection.

3.11: Which two kinds of selection are employed in the process of Man's Selection?

3.12: Into what type of selection does Unconscious Selection graduate?

3.13: According to Darwin, how might Man's Selection and Natural Selection interact?

> *During this investigation we shall see that the principle of Selection is highly important. Although man does not cause variability and cannot even prevent it, he can select, preserve, and accumulate the variations given to him by the hand of nature almost in any way which he chooses; and thus he can certainly produce a great result. ... Man may select and preserve each successive variation, with the distinct intention of improving and altering a breed, in accordance with a preconceived idea; and by thus adding up variations, often so slight as to be imperceptible by an uneducated eye, he has effected wonderful changes and improvements.*
>
> Darwin, Charles. *Variation Under Domestication,* Page 4.
>
> *Over all these causes of Change, the accumulative action of Selection, whether applied methodically and quickly, or unconsciously and slowly but more efficiently, seems to have been the predominant Power.*
>
> Darwin, Charles. *The Origin of Species,* page 32

3.14: What did Darwin observe as the predominant Power in creation and evolution?

3.15: How did Darwin compare the action of methodical selection and unconscious selection?

3.16: Through advancements in genetic engineering, humans can now introduce individual variations to organisms at the gene level. How does this alter Darwin's observations regarding Man's Selection?

> *On the view here given of the important part which selection by man has played, it becomes at once obvious, how it is that our domestic races show adaptation in their structures or in their habits to man's wants or fancies.*
>
> Darwin, Charles. *The Origin of Species,* page 28

3.17: Man's Selection played an important part in Darwin's theory by making it obvious that Selection results in the adaptation of what two broad categories of traits or characteristics?

3.18: Describe several structural traits that humans may desire to preserve in a domestic species, and why?

3.19: Describe several habits, or behavioral traits, which humans may desire to preserve in domestic species, and why?

Answering the Inquiry: How do Darwin's observations address the Creation and Evolution of all living things?

Notes:

CHAPTER 4
Darwin on Natural Selection

Student Inquiry: Is Natural Selection the exclusive means of evolutionary modification, or is it one type of Selection which acts upon one or more types of Variation?

Can it, then, be thought improbable, seeing that variations useful to man have undoubtedly occurred, that other variations useful in some way to each being in the great and complex battle of life, should occur in the course of many successive generations? If such do occur, can we doubt (remembering that many more individuals are born than can possibly survive) that individuals having any advantage, however slight, over others, would have the best chance of surviving and of procreating their kind? On the other hand, we may feel sure that any variation in the least degree injurious would be rigidly destroyed. This preservation of favourable individual differences and variations, and the destruction of those which are injurious, I have called Natural Selection, or the Survival of the Fittest.

Darwin, Charles. *The Origin of Species*, page 63

But if variations useful to any organic being ever do occur, assuredly individuals thus characterised will have the best chance of being preserved in the struggle for life; and from the strong principle of inheritance, these will tend to produce offspring similarly characterised. This principle of preservation, or the survival of the fittest, I have called Natural Selection.

Darwin, Charles. *The Origin of Species*, page 103

For brevity sake I sometimes speak of natural selection as an intelligent power;- in the same way as astronomers speak of the attraction of gravity as ruling the movements of the planets, or as agriculturists speak of man making domestic races by his power of selection. In the one case, as in the other, selection does nothing without variability, and this depends in some manner on the action of the surrounding circumstances of the organism.

Darwin, Charles *Variation Under Domestication*, page 6

4.1: How did Darwin define Natural Selection?

4.2: Compare and contrast the driving force of Natural Selection with the driving force of Man's Selection.

4.3: Darwin sometimes spoke of natural selection as what type of power? What two analogies did he draw with regard to this power?

> *We have seen that man by selection can certainly produce great results, and can adapt organic beings to his own uses, through the accumulation of slight but useful variations, given to him by the hand of Nature. But Natural Selection, we shall hereafter see, is a power incessantly ready for action, and is as immeasurably superior to man's feeble efforts, as the works of Nature are to those of Art.*
>
> Darwin, Charles. *The Origin of Species*, page 49.
>
> *Natural Selection acts exclusively by the preservation and accumulation of variations, which are beneficial under the organic and inorganic conditions to which each creature is exposed at all periods of life. The ultimate result is that each creature tends to become more and more improved in relation to its conditions. This improvement inevitably leads to the gradual advancement of the organisation of the greater number of living beings throughout the world. But here we enter on a very intricate subject, for naturalists have not defined to each other's satisfaction what is meant by advance in organisation.*
>
> Darwin, Charles. *The Origin of Species*, page 97

4.4: What types of variations in plants and animals are preserved and accumulated under Natural Selection, and what is the determining factor?

4.5: What types of variation in plants and animals are preserved and accumulated under Man's Selection, and what is the determining factor?

4.6: Give examples of structural and behavioral traits selected by man for a domestic species that would not be preserved if that species was left unprotected (competing to survive) and subject solely to Natural Selection?

> *In the second place, we may easily err in attributing importance to characters, and in believing that they have been developed through natural selection. We must by no means overlook the effects of the definite action of changed conditions of life, - of so-called spontaneous variations, which seem to depend in a quite subordinate degree on the nature of the conditions, - of the tendency to reversion to long-lost characters, - of the complex laws of growth, such as of correlation, compensation, of the pressure of one part on another, etc., - and finally of sexual selection, by which characters of use to one sex are often gained and then transmitted more or less perfectly to the other sex, though of no use to the sex.*
>
> Darwin, Charles. *The Origin of Species*, page 158

4.7: Darwin recognized that some characteristics may not have been developed through Natural Selection. What other processes did Darwin observe may effect changes in species outside the effect of Natural Selection?

> *It may be doubted whether sudden and considerable deviations of structure, such as we occasionally see in our domestic productions, more especially with plants, are ever permanently propagated in a state of nature.*
>
> Darwin, Charles. *The Origin of Species*, page 33
>
> *I am inclined to suspect that we see, at least in some of these polymorphic genera, variations which are of no service or disservice to the species, and which consequently have not been seized on and rendered definite by natural selection, as hereafter to be explained.*
>
> Darwin, Charles. *The Origin of Species*, page 35
>
> *Variations neither useful nor injurious would not be affected by natural selection, and would be left either a fluctuating element, as perhaps we see in certain polymorphic species, or would ultimately become fixed, owing to the nature of the organism and the nature of the conditions.*
>
> Darwin, Charles. *The Origin of Species*, page 63
>
> *Some writers have misapprehended or objected to the term Natural Selection. Some have even imagined that natural selection induces variability, whereas it implies only the preservation of such variations as arise and are beneficial to the being under its conditions of life.*
>
> Darwin, Charles. *The Origin of Species*, page 63.
>
> *That natural selection generally acts with extreme slowness I fully admit.*
>
> Darwin, Charles. *The Origin of Species*, page 84.

4.8: Darwin found that the occurrence of some traits in species could not be explained through Natural Selection. What did Darwin observe regarding the action of Natural Selection on the following:

4.8(a): Sudden changes in the structure of a species?

4.8(b): Polymorphic traits, such as eye color?

4.8(c): The rate of species modification?

4.8(d): The inducement of variations?

> *It is scarcely possible to avoid comparing the eye with a telescope. We know that this instrument has been perfected by the long-continued efforts of the highest human intellects; and we naturally infer that the eye has been formed by a somewhat analogous process. But may not this inference be presumptuous? Have we any right to assume that the Creator works by intellectual powers like those of man?*
>
> Darwin, Charles. *The Origin of Species*, page 146

> *Throughout this chapter and elsewhere I have spoken of selection as the paramount power, yet its action absolutely depends on what we in our ignorance call spontaneous or accidental variability. Let an architect be compelled to build an edifice with uncut stones, fallen from a precipice. The shape of each fragment may be called accidental; yet the shape of each has been determined by the force of gravity, the nature of the rock, and the slope of the precipice, - events and circumstances, all of which depend on natural laws; but there is no relation between these laws and the purpose for which each fragment is used by the builder. In the same manner the variations of each creature are determined by fixed and immutable laws; but these bear no relation to the living structure which is slowly built up through the power of selection, whether this be natural or artificial selection.*
>
> *If our architect succeeded in rearing a noble edifice, using the rough wedge-shaped fragments for the arches, the longer stones for the lintels, and so forth, we should admire his skill even in a higher degree than if he had used stones shaped for the purpose. So it is with selection, whether applied by man or nature; for although variability is indispensably necessary, yet, when we look at some highly complex and excellently adapted organism, variability sinks to a quite subordinate position in importance in comparison with selection, in the same manner as the shape of each fragment used by our supposed architect is unimportant in comparison with his skill.*
>
> Darwin, Charles. *Variation Under Domestication*, Volume II page 236.

4.9: Compare and contrast the design and use of the diverse eye types observable in nature (e.g., pit eye, compound eye, binocular eye, etc.) with the design and use of a telescope?

4.10: Do you think comparing the eye with a telescope is a fair comparison?

4.11: Do you agree with Darwin that it may be presumptions to assume that *"the Creator"* uses intellectual powers the same way man does?

4.12: Try to match the elements of the Architect analogy with the Creation and Evolution system diagram on page 16.

4.12(a): What represents Variation in the above analogy?

4.12(b): What causes the Variations?

4.12(c): What represents Selection in the above analogy?

4.12(d): What is the driving force behind the Selection?

Answering the Inquiry: Is Natural Selection the exclusive means of evolutionary modification, or is it one type of Selection which acts upon one or more types of Variation?

Notes:

CHAPTER 5
Darwin on Difficulties with the Theory of Natural Selection

Student Inquiry: Does the scientific evidence of Revolutionary Evolution conflict with the theory of Natural Selection?

Abrupt or sudden changes in species and the rapid modification or development of species are referred to herein as **Revolutionary Evolution**. Darwin was concerned that the evidence of such changes created serious difficulties for the theory of Natural Selection as the sole means of evolutionary change.

> *Long before the reader has arrived at this part of my work, a crowd of difficulties will have occurred to him. Some of them are so serious that to this day I can hardly reflect on them without being in some degree staggered; but, to the best of my judgment, the number are only apparent, and those that are real are greater not, I think, fatal to the theory.*
>
> Darwin, Charles. *The Origin of Species*, page 133
>
> *As natural selection acts solely by accumulating slight, successive, favourable variations, it can produce no great or sudden modifications; it can act only by short and slow steps.*
>
> Darwin, Charles. *The Origin of Species*, page 413
>
> *The abrupt manner in which whole groups of species suddenly appear in certain formations, has been urged by several palaeontologists - for instance, by Agassiz, Pictet, and Sedgwick - as a fatal objection to the belief in the transmutation of species. If numerous species, belonging to the same genera or families, have really started into life at once, the fact would be fatal to the theory of evolution through natural selection.*
>
> Darwin, Charles. *The Origin of Species*, page 282

5.1: Review questions 4.7(a)-(d). Identify some of the difficulties Darwin observed with the exclusivity of Natural Selection.

5.2: Can abrupt changes in species, or Revolutionary Evolution, be attributed to Natural Selection?

5.3: Can the mass extinction of species be attributed to Natural Selection?

> *On Extinction*
> *We have as yet spoken only incidentally of the disappearance of species and of groups of species. On the theory of natural selection, the extinction of old forms and the production of new and improved forms are intimately connected together.*
> Darwin, Charles. *The Origin of Species*, page 293
>
> *On this doctrine of the extermination of an infinitude of connecting links, between the living and extinct inhabitants of the world, and at each successive period between the extinct and still older species, why is not every geological formation charged with such links? Why does not every collection of fossil remains afford plain evidence of the gradation and mutation of the forms of life? Although geological research has undoubtedly revealed the former existence of many links, bringing numerous forms of life much closer together, it does not yield the infinitely many fine gradations between past and present species required on the theory, and this is the most obvious of the many objections which may be urged against it.*
> Darwin, Charles. *The Origin of Species*, page 407

5.4: What did Darwin observe as the most obvious of the many objections to the exclusivity of the theory of Natural Selection?

> *To the question why we do not find rich fossiliferous deposits belonging to these assumed earliest periods prior to the Cambrian system, I can give no satisfactory answer.*
> Darwin, Charles. *The Origin of Species*, page 286
>
> *The several difficulties here discussed, namely, that, though we find in our geological formations many links between the species which now exist and which formerly existed, we do not find infinitely numerous fine transitional forms closely joining them all together. The sudden manner in which several groups of species first appear in our European formations, the almost entire absence, as at present known, of formations rich in fossils beneath the Cambrian strata, are all undoubtedly of the most serious nature.*
> Darwin, Charles. *The Origin of Species*, page 289

Conduct research on the Cambrian Explosion.

5.5: When do scientists believe the Cambrian explosion occurred?

5.6: Considering the Creation and Evolution System Diagram on page 16 and the notes you've taken thus far, what combination of Variation and Selection processes could explain the Cambrian explosion?

> *Why then is not every geological formation and every stratum full of such intermediate links? Geology assuredly does not reveal any such finely-graduated organic chain; and this, perhaps, is the most obvious and serious objection which can be urged against the theory, The explanation lies, as I believe, in the extreme imperfection of the geological record.*
>
> Darwin, Charles. *The Origin of Species*, pages 264 - 265
>
> *Although we now know that organic beings appeared on this globe, at a period incalculably remote, long before the lowest bed of the Cambrian system was deposited, why do we not find beneath this system great piles of strata stored with the remains of the progenitors of the Cambrian fossils? For on the theory, such strata must somewhere have been deposited at these ancient and utterly unknown epochs of the world's history.*
>
> *I can answer these questions and objections only on the supposition that the geological record is far more imperfect than most geologists believe.*
>
> Darwin, Charles. *The Origin of Species*, page 408

5.7: Darwin observed that the geological record reveals the most obvious and serious objection to the theory of Natural Selection. What objection does the geological record raise?

5.8: How did Darwin explain the lack of geological evidence supporting the theory of Natural Selection?

5.9: According to Darwin, what must exist beneath the Cambrian strata (in other words, have existed before the Cambrian period) in order to support the exclusivity of Natural Selection?

5.10: The lack of fossil evidence prior to the Cambrian period indicates the rapid creation and evolution of numerous species during the Cambrian revolution. How did Darwin explain the inconsistency between the fossil evidence prior to the Cambrian period and the theory of Natural Selection?

5.11: Considering what you now know about Man's Selection, do you think this process can drive the creation and evolution of a species more rapidly than Natural Selection?

5.12: Recall from Chapter 2 that Variation is the catalyst for Selection. Review previous chapters and consider if there are forms of Variation that may cause more rapid advancement in the development of a species than Individual Variation.

5.13: Do you think that different forms of Variation may be combined with different forms of Selection to create periods of Revolutionary Evolution. Why or why not?

> *When discussing special cases, Mr. Mivart passes over the effects of the increased use and disuse of parts, which I have always maintained to be highly important, and have treated in my "Variation under Domestication" at greater length than, as I believe, any other writer. He likewise often assumes that I attribute nothing to variation, independently of natural selection, whereas in the work just referred to I have collected a greater number of well-established cases than can be found in any other work known to me.*
>
> Darwin, Charles. *The Origin of Species*, page 176
>
> *It may be doubted whether sudden and considerable deviations of structure, such as we occasionally see in our domestic productions, more especially with plants, are ever permanently propagated in a state of nature.*
>
> Darwin, Charles. *The Origin of Species*, page 33
>
> *But we learn from the study of our domestic productions that the disuse of parts leads to their reduced size; and that the result is inherited.*
>
> Darwin, Charles. *The Origin of Species*, page 400
>
> *Under domestication we see much variability, caused, or at least excited, by changed conditions of life; but often in so obscure a manner, that we are tempted to consider the variations as spontaneous.*
>
> Darwin, Charles. *The Origin of Species*, page 410
>
> *Furthermore, I am convinced that Natural Selection has been the most important, but not the exclusive, means of modification.*
>
> Darwin, Charles. *The Origin of Species*, page 4

5.14: Where did Darwin occasionally observe examples of sudden or considerable deviations in structure, or Revolutionary Evolution?

5.15: Under domestication we see much variability as a result of what actions?

5.16: If Natural Selection is not the exclusive means of modification, and if other combinations of Variation and Selection may result in Revolutionary Evolution as seen in domestic species, are Revolutionary Evolution and Natural Selection mutually exclusive?

Answering the Inquiry: Does the scientific evidence of Revolutionary Evolution conflict with the theory of Natural Selection?

Notes:

CHAPTER 6
Darwin on Domestication

Student Inquiry: Can purposeful creations be produced by Nature through a process analogous to Man's Selection?

> *Can the principle of selection, which we have seen is so potent in the hands of man, apply under nature? I think we shall see that it can act most efficiently.*
>
> Darwin Charles, *The Origin of Species,* page 62
>
> *It is scarcely possible to doubt that the long-continued selection of qualities serviceable to man has been the chief agent in the formation of the several breeds of horses. ... We may therefore conclude that, whether or not the various existing breeds of the horse have proceeded from one or more aboriginal stocks, yet that a great amount of change has resulted from the direct action of the conditions of life, and probably a still greater amount from the long-continued selection by man of slight individual differences.*
>
> Darwin, Charles. *Variation Under Domestication*, Volume I, page 56.

6.1: What two things have resulted in the great amount of change leading to the many breeds of the domestic horse?

6.2: What is the chief agent of change leading to the creation of the existing breeds of the domestic horse?

Look back at the process diagram on page 15 to discuss the following:

6.3(a): What is the primary selection process for the creation of a wild plant or animal species?

6.3(b): What is the driving force in the process in 6.3(a)?

6.4(a): What is the primary selection process for the creation of a domestic plant or animal species?

6.4(b): What is the driving force in the process in 6.4(a)?

> *As the will of man thus comes into play, we can understand how it is that domesticated breeds show adaptation to his wants and pleasures. We can further understand how it is that domestic races of animals and cultivated races of plants often exhibit an abnormal character, as compared with natural species; for they have been modified not for their own benefit, but for that of man.*
>
> Darwin, Charles. *Variation Under Domestication*, Volume I, page 4.

> *But we cannot explain by crossing the origin of such extreme forms as thoroughbred greyhounds, bloodhounds, bulldogs, Blenheim spaniels, terriers, pugs, &c., unless we believe that forms equally or more strongly characterized in these different respects existed in nature. But hardly any one has been bold enough to suppose that such unnatural forms ever did or could exist in a wild state. When compared with all known members of the family of Canidae they betray a distinct and abnormal origin. No instance is on record of such dogs as bloodhounds, spaniels, true greyhounds having been kept by savages: they are a product of long-continued civilization.*
>
> Darwin, Charles. *Variation Under Domestication*, Volume 1, page 35.

6.5: Did Darwin believe that the domestic dog could have been created simply by taming a wild progenitor species?

6.6: From the previous two passages, what types of characters or forms does the process of domestication produce, and for what purpose are such adaptations made?

Darwin studied the creation of the domestic dog at length as part of his analysis of Man's Selection and domestication.

> *The first and chief point of interest in this chapter is, whether the numerous domesticated varieties of the dog have descended from a single wild species, or from several. Some authors believe that all have descended from the wolf, or from the jackal, or from an unknown and extinct species. Others again believe, and this of late has been the favourite tenet, that they have descended from several species, extinct and recent, more or less commingled together. We shall probably never be able to ascertain their origin with certainty.*
>
> Darwin, Charles, *Variation Under Domestication*, page 15
>
> *It is notorious how greatly the mental disposition, tastes, habits, consensual movement, loquacity or silence, and tone of voice have varied and been inherited in our domestic animals. The dog offers the most striking instance of changed mental attributes, and these differences cannot be accounted for by descent from distinct wild types.*
>
> Darwin, Charles, *Variation Under Domestication*, Volume II, page 404

6.7: Based on Darwin's observation, do you believe that the domestic dog can be linked to a single progenitor species?

6.8: The domestic dog demonstrates the most striking example of what inherited mental attributes?

6.9: According to Darwin, what process cannot account for the differences in mental attributes between wild and domestic canines?

6.10: Research the origin of the domestic dog and decide for yourself when you think the domestic dog was created.

> *Let us now look to the action of natural selection on special characters. Although nature is difficult to resist, yet man often strives against her power, and sometimes with success. From the facts to be given, it will also be seen that natural selection would powerfully affect many of our domestic productions if left unprotected.*
>
> Darwin, Charles. *Variation Under Domestication* (1875), Volume II, Page 211.
>
> *As a consequence of continued variability, and more especially of reversion, all highly improved races, if neglected or not subjected to incessant selection, soon degenerate.*
>
> Darwin, Charles. *Variation Under Domestication*, Volume II, page 225.

6.11: How does the domestication of the dog demonstrate human success in overcoming the power of Natural Selection (descent with modification) using the process of Man's Selection?

6.12: What happens to a domestic species if left unprotected by its creator and therefore subject only to Natural Selection?

> *We here see that there is no need to separate single pairs, as man does, when he methodically improves a breed: natural selection will preserve and thus separate all the superior individuals, allowing them freely to intercross, and will destroy all the inferior individuals. By this process long-continued, which exactly corresponds with what I have called unconscious selection by man, combined no-doubt in a most important manner with the inherited effects of the increased use of parts, it seems to me almost certain that an ordinary hoofed quadruped might be converted into a giraffe.*
>
> Darwin, Charles. *The Origin of Species*, pages 177 - 178,

6.13: According to Darwin, the process of Natural Selection, combined with the inherited effects of the increased use of parts (see IAC Variations, Chapter 2), exactly corresponds with what process?

6.14: What wild species was Darwin almost certain may have resulted from such a process?

6.15: Unconscious Selection graduates into what process (see page 27)?

> *Again, we may suppose that at an early period of history, the men of one nation or district required swifter horses, whilst those of another required stronger and bulkier horses. The early differences would be very slight; but in the course of time, from the continued selection of swifter horses in the one case, and of stronger ones in the other, the differences would become greater, and would be noted as forming two sub-breeds. Ultimately, after the lapse of centuries, these sub-breeds would become converted into two well-established and distinct breeds. As the differences become greater, the inferior animals with intermediate characters, being neither very swift nor very strong, would not have been used for breeding, and will thus have tended to disappear. Here, then, we see in man's productions the action of what may be called the principle of divergence, causing differences, at first barely appreciable, steadily to increase, and the breeds to diverge in character, both from each other and from their common parent.*
>
> *But how it may be asked, can any analogous principle apply in nature? I believe it can and does ally most efficiently (though it was a long time before I saw how), from the simple circumstances that the more diversified the descendants from any one species become in structure, constitution, and habits, by so much will they be better enabled to seize on many and widely diversified places in the polity of nature, ...*
>
> Darwin, Charles. *The Origin of Species,* page 87
>
> *An unexplained residuum of change must be left to the assumed uniform action of those unknown agencies, which occasionally induce strongly marked and abrupt deviations of structure in our domestic productions.*
>
> Darwin, Charles. *The Descent of Man,* page 62.

6.16: Is the process used in the creation of divergent breeds of horse, as described by Darwin, purposeful?

6.17: According to Darwin, can an analogous principle apply in nature?

6.18: What agencies other than man did Darwin observe may induce strongly marked and abrupt deviations in domestic productions?

6.19: Considering the processes of Variation and Selection, what types of actions might these agencies take to induce such deviations?

6.20: Based on Darwin's conviction of the high value of the study of variation under domestication to gain knowledge on perplexing cases (refer to Darwin's quote on page 13), do you think that principles and agencies analogous to those observed under domestication occur in nature?

Answering the Inquiry: Can purposeful creations be produced by Nature through a process analogous to Man's Selection?

Notes:

CHAPTER 7
Evolution: Reducing Humans

Student Inquiry: Can the creation and evolution of modern humans be attributed exclusively to Natural Selection?

> *In consequence of the views now adopted by most naturalists, and which will ultimately, as in every other case, be followed by others who are not scientific, I have been led to put together my notes, so as to see how far the general conclusions arrived at in my former works were applicable to man. This seemed all the more desirable, as I have never deliberately applied these views to a species taken singly.*
>
> *... The sole object of this work is to consider, firstly, whether man, like every other species, is descended from some preexisting form; secondly, the manner of his development; and thirdly, the value of the differences between the so-called races of man.*
>
> Darwin, Charles. *The Descent of Man,* pages 1 - 2.
>
> *Many of the views which have been advanced are highly speculative, and some no doubt will prove erroneous; but I have in every case given reasons which have led me to one view rather than to another. It seemed worth while to try how far the principle of evolution would throw light on some of the more complex problems in the natural history of man. False facts are highly injurious to the progress of science, for they often endure long; but false views, if supported by some evidence, do little harm, for every one takes a salutary pleasure in proving their falseness; and when this is done, one path towards error is closed and the road to truth is often at the same time opened.*
>
> Darwin, Charles. *The Descent of Man,* page 606

In his work entitled *The Descent of Man,* Darwin attempted to see *"how far the principle of evolution would throw light on some of the more complex problems in the natural history of man."* Darwin advanced *"highly speculative views,"* some of which would *"no doubt prove erroneous,"* to determine if the creation of modern humans could be reduced to the next step in primate evolution exclusively through the process of Natural Selection.

7.1: What were Darwin's three objectives for *The Descent of Man?*

7.2: Why do you think Darwin would test his theory of Natural Selection by advancing highly speculative and possibly erroneous views?

7.3: What did Darwin expect would happen to any false views that he advanced?

7.4: How did Darwin think that false views on creation and evolution, if advanced as facts, could impact the progress of science?

Thomas Huxley was referred to as Darwin's Bulldog. He strongly advocated Natural Selection as the exclusive means of evolution and creation of all species. Based on the anthropological analysis of primates, Huxley was adamant that humans were merely an advanced form of ape. Darwin relied heavily on Huxley's opinions in advancing his views. Huxley was also the founder of the X-Club, a liberal dining club devoted to naturalism and the removal of the concept of a Creator from science education. Research Thomas Huxley and the X-Club.

> *Nor shall I have occasion to do more than to allude to the amount of difference between man and the anthropomorphous apes; for Prof. Huxley, in the opinion of most competent judges, has conclusively shewn that in every visible character man differs less from the higher apes, than these do from the lower members of the same order of Primates.*
>
> Darwin, Charles. *The Descent of Man*, page 2.
>
> *It would be beyond my limits, and quite beyond my knowledge, even to name the innumerable points of structure which man agrees with the other Primates. Our great anatomist and philosopher, Prof. Huxley, has fully discussed this subject, and concludes that man in all parts of his organization differs less from the higher apes, than these do from the lower members of the same group. Consequently, there "is not justification for placing man in a distinct order."*
>
> Darwin, Charles. *The Descent of Man*, page 150.

7.5: Do you think Huxley's beliefs may have influenced the views he presented as facts regarding the difference between man and higher apes?

> *Looking to future generations, there is no cause to fear that the social instincts will grow weaker, and we may expect that virtuous habits will grow stronger, becoming perhaps fixed by inheritance. In this case the struggle between our higher and lower impulses will be less severe, and virtue will be triumphant.*
>
> Darwin, Charles. *The Descent of Man*, page 125

7.6: In the above passage Darwin speculated on the future of humanity. Considering the events of the 20th century, do you think human social instincts naturally cause virtuous habits (e.g., morality, ethics, altruism, civility) to *"grow stronger"*, and that these habits will be *"fixed by inheritance"*?

Summary of the last two Chapters.- There can be no doubt that the difference between the mind of the lowest man and that of the highest animal is immense. An anthropomorphous ape, if he could take a dispassionate view of his own case, would admit that though he could form an artful plan to plunder a garden - though he could use stones for fighting off or breaking open nuts, yet that the thought of fashioning a stone into a tool was quite beyond his scope. Still less, as he would admit, could he follow out a train of metaphysical reasoning, or solve a mathematical problem, or reflect on God, or admire a grand natural scene. Some apes, however, would probably declare that they could and did admire the beauty of the coloured skin and fur of their partners in marriage. They would admit, that though they could make other apes understand by cries some of their perceptions and simpler wants, the notion of expressing definite ideas by definite sounds had never crossed their minds. They might insist that they were ready to aid their fellow-apes of the same troop in many ways, to risk their lives for them, and to take charge of their orphans; but they would be forced to acknowledge that disinterested love for all living creatures, the most noble attribute of man, was quite beyond their comprehension.

Nevertheless the differences in mind between man and higher animals, great as it is, certainly is one of degree and not of kind.

Darwin, Charles. *The Descent of Man*, page 125

Even if it be granted that the difference between man and his nearest allies is as great in corporeal structure as some naturalists maintain, and although we must grant that the difference between them is immense in mental power, yet the facts given in the earlier chapters appear to declare, in the plainest manner, that man is descended from lower form, notwithstanding that connecting-links have not hitherto been discovered.

Darwin, Charles. *The Descent of Man*, page 146.

7.7: Discuss Darwin's comparison of the mental abilities of the anthropomorphous ape with that of modern humans in the following areas: (1) forming tools, (2) metaphysical reasoning, (3) solving a mathematical problem, (4) reflecting on God, (5) admiration, (6) expressing ideas through language, and (7) unconditional love.

7.8: What did Darwin assert was the most notable attribute of humans?

7.9: What was Darwin's view regarding the great difference in mind between humans and higher animals? Do you think Darwin's view is a fact?

7.10: Could Spontaneous Variation (see Chapter 2) explain the abrupt creation of human mental faculties without leaving any connecting links?

> *Thus a large yet undefined extension may be given to the direct and indirect results of natural selection; but I now admit, after reading the essay by Nageli on plants, and the remarks by various authors with respect to animals, more especially those recently made by Professor Broca, that in the earlier editions of my 'Origin of Species' I perhaps attributed too much to the action of natural selection or the survival of the fittest. ... Any one with this assumption in his mind would naturally extend too far the action of natural selection, either during past or present times. Some of those who admit the principle of evolution, but reject natural selection, seem to forget, when criticising my book, that I had the above two objects in my view; hence if I have erred in giving natural selection great power, which I am very far from admitting, or in having exaggerated its power, which is in itself probable, I have at least, as I hope, done good service in aiding to overthrow the dogma of separate creations.*
>
> Darwin, Charles. *The Descent of Man*, page 61.

7.11: What did Darwin admit was *"probable"* with regard to his observations on the power of Natural Selection in *The Descent of Man*?

7.12: Research and describe Nageli's theory of the Inner Perfecting Principle.

> *On the view here given of the important part which selection by man has played, it becomes at once obvious, how it is that our domestic races show adaptation in their structures or in their habits to man's wants or fancies?*
>
> Darwin, Charles. *The Origin of Species*, page 28
>
> *Important as the struggle for existence has been and even still is, yet as far as the highest part of man's nature is concerned there are other agencies more important. For the moral qualities are advanced, either directly or indirectly, much more through the effects of habit, the reasoning powers, instruction, religion, &c., than through natural selection; though to this latter agency may be safely attributed the social instincts, which afforded the basis for the development of the moral sense.*
>
> Darwin, Charles, *The Descent of Man*, page 618.

7.13: Darwin believed that social instincts are the basis for moral behavior. Is there a difference between social behavior and moral behavior?

7.14: As far as the highest part of man's nature is concerned, what agencies did Darwin observe are more important in advancing the moral qualities than Natural Selection ?

Answer the Inquiry: Can the creation and evolution of modern humans be attributed exclusively to Natural Selection?

Notes:

CHAPTER 8
Creation: Elevating Humanity

Student Inquiry: Could modern humans be a domestic species purposefully created through a process analogous to Man's Selection?

> *I fully subscribe to the judgment of those writers who maintain that of all the differences between man and the lower animals, the moral sense of conscience is by far the most important. This sense, as Mackintosh remarks, "has a rightful supremacy over every other principle of human action;" it is summed up in that short but imperious word 'ought,' so full of high significance. It is the most noble of all the attributes of man, leading him without a moment's hesitation to risk his life for that of a fellow-creature; or after due deliberation, impelled simply by the deep feeling of right or duty, to sacrifice it in some great cause. ...*
>
> *... The investigation possesses, also, some independent interest, as an attempt to see how far the study of the lower animals throws light on one of the highest psychical faculties of man.*
>
> *The following proposition seems to me in a high degree probable - namely, that any animal whatever, endowed with well-marked social instincts, the parental and filial affections being here included, would inevitably acquire a moral sense of conscience, as soon as its intellectual powers had become as well, or nearly as well developed, as in man.*
>
> Darwin, Charles. *The Descent of Man,* page 97, 98.
>
> *Man in the rudest state in which he now exists is the most dominant animal that has ever appeared on earth. He has spread more widely that(sic) any other highly organised form: and all others have yielded before him. He manifestly owes this immense superiority to his intellectual faculties, to his social habits, which lead him to aid and defend his fellows, and to his corporeal structure.*
>
> Darwin, Charles. *The Descent of Man,* page 45.

8.1: What did Darwin observe was the most important difference between man and the lower animals? Do you think this makes humans a singular species?

8.2: According to Darwin, what would be required for an animal to acquire a moral sense of conscience?

8.3: To what do humans owe their immense superiority over all other animals?

> *A moral being is one who is capable of comparing his past and future actions or motives, and approving or disapproving of them. We have no reason to suppose that any of the lower animals have this capacity; therefore, when a Newfoundland dog drags a child out of the water, or a monkey faces danger to rescue its comrade, or takes charge of an orphan monkey, we do not call its conduct moral. But in the case of man, who alone can with certainty be ranked as a moral being, actions of a certain class are called moral, whether performed deliberately, after a struggle with opposing motives, or impulsively through instinct, or from the effects of slowly-gained habit.*
>
> Darwin, Charles. *The Descent of Man*, page 111.

Darwin accepted Huxley's anthropological views regarding the connection between humans and apes as facts. Yet he could not explain through Natural Selection alone the singularity of human morality, "*one of the highest psychical faculties of man.*" Human morality cannot be observed in any wild species, and therefore is singular to human intelligence.

8.4: How did Darwin differentiate the actions or motives of a moral being? Did he think that any animal, other than humans, have this capacity?

8.5: Name some moral human behaviors that are singular to human intelligence? We will define these moral qualities as **"Singular Intelligence."**

Darwin observed that while wild animals do not demonstrate moral behavior, we can see such qualities in some domestic species.

> *Our domestic dogs are descended from wolves and jackals, and though they may not have gained in cunning, and may have lost in wariness and suspicion, yet they have progressed in certain moral qualities, such as in affection, trustworthiness, temper, and probably in general intelligence.*
>
> Darwin, Charles. *The Descent of Man*, page 80.
>
> *The feeling of religious devotion is a highly complex one, consisting of love, complete submission to an exalted and mysterious superior, a strong sense of dependence, fear, reverence, gratitude, hope for the future, and perhaps other elements. No being could experience so complex an emotion until advanced in his intellectual and moral faculties to at least a moderately high level. Nevertheless, we see some distant approach to this state of mind in the deep love of a dog for his master, associated with complete submission, some fear, and perhaps other feelings. ... Professor Braubach goes so far as to maintain that a dog looks on his master as on a god.*
>
> Darwin, Charles. *The Descent of Man*, page 95 - 96

8.6: Darwin observed moral qualities in what domestic species?

8.7: Do you think man has purposefully instilled some of his own moral qualities in the domestic dog through the process of Man's Selection?

8.8: What traits did Darwin observe that domestic dogs may have lost through the process of Man's Selection?

8.9: To what purpose do you think humans would supplant survival instincts with moral instincts in domestic dogs?

Alfred Russel Wallace co-discovered the theory of Natural Selection through scientific work and publication contemporaneous with Darwin's. Darwin referenced Wallace's observations on numerous occasions and they agreed on many aspects of Natural Selection. However, Darwin disagreed with Wallace's assertion that Natural Selection could only have produced a human mind slightly superior to an ape, and therefore could not explain modern human behavior even in an uncivilized or "savage" state. Wallace further asserted that modern humans were *"a new and distinct order of being;"* ... *"not by a change in body, but by an advance of mind."*

These several inventions, by which man in the rudest state has become so preeminent, are the direct result of the development of his powers of observation, memory, curiosity, imagination, and reason. I cannot, therefore, understand how it is that Mr. Wallace maintains, that "natural selection could only have endowed the savage with a brain a little superior to that of an ape."

Darwin, Charles. *The Descent of Man*, page 138.

If the views I have here endeavored to sustain have any foundation, they give us a new argument for placing man apart, as not only the head and culminating point of the grand series of organic nature, but as in some degree a new and distinct order of being. ...

*At length, however, there came into existence a being in whom that subtle force we term **mind,** became of greater importance than his mere bodily structure. ... - a being who was in some degree superior to nature, inasmuch, as he knew how to control and regulate her action, and could keep himself in harmony with her, not by a change in body, but by an advance of mind.*

Here, then, we see the true grandeur and dignity of man. ... Man has not only escaped "natural selection" himself, but he actually is able to take away some of that power from nature which, before his appearance, she universally exercised.

Wallace, Alfred R. *"The Origin of Human Races and the Antiquity of Man Deduced from the Theory of 'Natural Selection'"*, Journal of the Anthropological Society, (1864).

Wallace observed that modern man was created, or *"came into existence"* when the *"subtle force we term mind, became more important than his mere bodily structure."* This concept may be defined as **Behavioral Modernity.**

8.10: Research and discuss the term "Behavioral Modernity".

8.11: Do you think that the human body (structure) or the human mind (behavior) is more important in defining modern humans?

Now no animal can be compared to man in this respect, for he is omnivorous, and dwells in every climate, and is far more domesticated and far more advanced from his first beginnings than any other animal;

Blumenbach, Johann Friedrich. *Treatise on Anthropology,' Eng. translat.,* 1865, p.205

With respect to the causes of variability, we are in all cases very ignorant; but we can see that in man as in the lower animals, they stand in some relation to the conditions to which each species has been exposed, during several generations. Domesticated animals vary more than those in a state of nature; and this is apparently due to the diversified and changing nature of the conditions to which they have been subjected. In this respect the different races of man resemble domesticated animals, and so do the individuals of the same race, when inhabiting a very wide area, like that of America. ... It is, nevertheless, an error to speak of man, even if we look only to the conditions to which he has been exposed, as "far more domesticated" [12] *than any other animal. ... .*

[12] Blumenbach, *'Treatise on Anthropology,' Eng. translat.,* 1865, p.205.

Darwin, Charles. *The Descent of Man,* page 28.

We might, therefore, expect that civilized men, who in one sense are highly domesticated, would be more prolific than wild men.

Darwin, Charles. *The Descent of Man,* page 45.

Man in many respects may be compared with those animals which have long been domesticated, and a large body of evidence can be advanced in favuor of the Pallasian doctrine.

Darwin, Charles. *The Descent of Man,* page 172.

Darwin recognized that human moral behavior may be reflected in our closest domestic animals. He also observed that some human characteristics are similar to those seen in domestic species. Based on his assertion that nature can do what man can do, consider this chapter's inquiry.

Answering the Inquiry: Could modern humans be a domestic species purposefully created through a process analogous to Man's Selection?

Notes:

CHAPTER 9
Intelligent Design: The Force of Mind

Student Inquiry: Are Singular Intelligence and the Force of Mind evidence of the Theory of Intelligent Design?

> [27] *... To do good in return for evil, to love your enemy, is a height of morality to which it may be doubted whether the social instincts would, by themselves, have ever led us. It is necessary that these instincts, together with sympathy, should have been highly cultivated and extended by the aid of reason, instruction, and the love or fear of God, before any such golden rule would be thought or obeyed.*
>
> Darwin, Charles. *The Descent of Man*, page 113.
>
> *The ennobling belief in God is not universal with man; and the belief in spiritual agencies naturally follows from other mental powers. The moral sense perhaps affords the best and highest distinction between man and the lower animals; but I need say nothing on this head, as I have so lately endeavored to shew that the social instincts,- the prime principle of man's moral constitution[50]- with the aid of active intellectual powers and the effects of habit, naturally lead to the golden rule, "As ye would that men should do to you, do ye to them likewise;" and this lies at the foundation of morality.* [50] *'The Thoughts of Marcus Aurelius,' &c., p.139*
>
> Darwin, Charles. *The Descent of Man*, page 126.

Darwin struggled to explain the development of the moral traits of modern human behavior, what we have defined as Singular Intelligence, exclusively through the process of Natural Selection. This struggle was evident in Darwin's observations regarding The Golden Rule. Review the last quote on page 52 and the above passages as you consider the following questions.

9.1: In the first passage, Darwin *"doubted whether the social instincts would, by themselves,"* have led to The Golden Rule. What other human traits were essential before such a rule would be *"thought or obeyed?"*

9.2: In the second passage, what did Darwin assert would *"naturally lead to the golden rule?"*

9.3: Do you think highly intelligent wild species with strong social tendencies, such as gorillas and dolphins, exhibit human moral qualities?

9.4: From what source did Darwin observe that human *"belief in spiritual agencies naturally follows ...?"*

> *Belief in God - Religion - There is no evidence that man was aboriginally endowed with the ennobling belief in the existence of an Omnipotent God. On the contrary there is ample evidence, derived not from hasty travellers [sic], but from men who have long resided with savages, that numerous races have existed, and still exist, who have no idea of one or more gods, and who have no words in their languages to express such an idea. The question is of course wholly distinct from the higher one, whether there exists a Creator and Ruler of the universe; and this has been answered in the affirmative by some of the highest intellects that have ever existed.*
>
> Darwin, Charles. *The Descent of Man*, page 93.
>
> *The belief in God has often been advanced as not only the greatest, but the most complete of all distinctions between man and the lower animals. It is however impossible, as we have seen, to maintain that this belief is innate or instinctive in man. On the other hand a belief in all-pervading spiritual agencies seems to be universal; and apparently follows from a considerable advance in man's reason, and from a still greater advance in his faculties of imagination, curiosity and wonder. I am aware that the assumed instinctive belief in God has been used by many persons as an argument for His existence. But this is a rash argument, as we should thus be compelled to believe in the existence of many cruel and malignant spirits, only a little more powerful than man; for the belief in them is far more general than in a beneficent Deity. The idea of a universal and beneficent Creator does not seem to arise in the mind of man until he has been elevated by long-continued culture.*
>
> Darwin, Charles. *The Descent of Man*, page 612.

Considering the above passages and the two passages on the previous page, what were Darwin's observations on the following:

9.5: Is human belief in *"all-pervading spiritual agencies"* universal?

9.6: Is the human *"love and fear of God"* important in the creation and adoption of the Golden Rule, the height of human morality?

9.7: How did he answer the question *"whether there exists a Creator?"*

9.8: Where and when does the idea of a *"universal and beneficent Creator"* arise?

9.9: Review the passage at the bottom of page 56. Do domestic dogs reflect some human moral traits? Compare the domestic dog's feelings toward its domesticating agent (humans) with human feelings toward God.

Next, consider Wallace's observations on the singularity of the human *"mind"* in the following passage, an excerpt of which was on page 57.

If the views I have here endeavored to sustain have any foundation, they give us a new argument for placing man apart, as not only the head and culminating point of the grand series of organic nature, but as in some degree a new and distinct order of being. From those infinitely remote ages, when the first rudiments of organic life appeared upon the earth, every plant, and every animal has been subject to one great law of physical change. ... No living thing could escape this law of its being; none could remain unchanged and live, amid the universal change around it.

*At length, however, there came into existence a being in whom that subtle force we term **mind,** became of greater importance than his mere bodily structure. Though with a naked and unprotected body, **this** gave him clothing against the varying inclemencies of the seasons. Though unable to compete with the deer in swiftness, or the wild bull in strength, **this** gave him weapons with which to capture or overcome both. Though less capable than most other animals of living on the herbs and the fruits that unaided nature supplies, this wonderful faculty taught him to govern and direct nature to his own benefit, and make her produce food for him when and where he pleased. From the moment when the first skin was used as a covering, when the first rude spear was formed to assist in the chase, the first seed sown or shoot planted, a grand revolution was effected in nature, a revolution which in all the previous ages of the earth's history had had no parallel, for a being had arisen who was no longer necessarily subject to change with the changing universe - a being who was in some degree superior to nature, inasmuch, as he knew how to control and regulate her action, and could keep himself in harmony with her, not by a change in body, but by an advance of mind.*

Here, then, we see the true grandeur and dignity of man. On this view of his special attributes, we may admit that even those who claim for him a position as an order, a class, or a sub-kingdom by himself, have some reason on their side. He is indeed a being apart, since he is not influenced by the great laws which irresistibly modify all other organic beings. Nay more: this victory which he has gained for himself gives him a directing influence over other existences. Man has not only escaped "natural selection" himself, but he actually is able to take away some of that power from nature which, before his appearance, she universally exercised.

Wallace, Alfred R. - reference citation shown on page 41.

Recall that Wallace is the co-discoverer of Natural Selection.

9.10: Wallace observed that the human force of *"mind"* was in some degree *"superior to nature."* Do you think that the human mind arose from nature or do you think it was created by a source *"superior to nature?"*

9.11: According to Wallace, from what have humans *"escaped?"*

> *Another source of conviction in the existence of God, connected with the reason and not with the feelings, impresses me as having much more weight. This follows from the extreme difficulty or rather impossibility of conceiving this immense and wonderful universe, including man with his capacity of looking far backwards and far into futurity, as the result of blind chance or necessity. When thus reflecting I feel compelled to look at a First Cause having an intelligent mind in some degree analogous to that of man; and I deserve to be called a Theist.*
>
> *This conclusion was strong in my mind about the time, as far as I can remember, when I wrote the Origin of Species; and it is since that time that is has very gradually with many fluctuations become weaker. ...*
>
> *I cannot pretend to throw the least light on such abstruse problems. The mystery of the beginning of all things is insoluble by us; and I for one must be content to remain an Agnostic.*
>
> Darwin, Charles. *The Autobiography of Charles Darwin 1809-1882,* With original omissions restored, Edited with Appendix and Notes by his grand-daughter Nora Barlow; Collins, St. James Place, London (1958) pages 92 - 93.
>
> *A grand and almost untrodden field of inquiry will be opened, on the causes and laws of variation, on correlation, on the effects of use and disuse, on the direct action of external conditions, and so forth. The study of domestic productions will rise immensely in value. A new variety raised by a man will be a more important and interesting subject for study than one more species added to the infinitude of already recorded species. Our classifications will come to be, as far as they can be so made, genealogies; and we will then truly give what may be called the plan of creation.*
>
> Darwin, Charles. *The Origin of Species,* page 427.

9.12: How did Darwin describe the *"First Cause"* of this *"immense and wonderful universe, including man"*?

9.13: Darwin asserted that it was impossible to conceive of the universe or humans as the result of what cause?

9.14: Where do you think Darwin's observations on creation and evolution fall with regard to the Case for Chance or the Case for Design?

9.15: Do you agree with Darwin that the study of Man's Selection may provide evidence revealing *"what may be called the plan of creation"*?

9.16: Do you think Darwin's views regarding a *"Creator,"* a *"First Cause,"* or a *"plan of creation"* are contrary to scientific observation?

Answering the Inquiry: Are Singular Intelligence and the Force of Mind evidence of the Theory of Intelligent Design?

Notes:

Using Your Singular Intelligence
SMART Learning

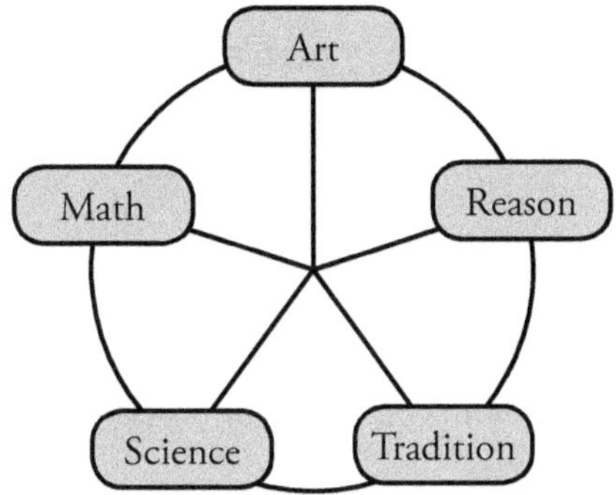

As a man who has devoted his whole life to the most clear headed science, to the study of matter, I can tell you as a result of my research about atoms this much: There is not matter as such. All matter originates and exists only by the virtue of a force which brings the particle of an atom to vibration and holds this most minute solar system of the atom together. We must assume behind this force the existence of a conscious and intelligent Mind. This Mind is the matrix of all matter.

Max Planck, *Das Wessen der Materie*, a lecture delivered in Florence Italy in 1944, excerpt in Gregg Braden in *The Spontaneous Healing of Belief; Shattering the Paradigm of False Limits* (2009), 334-335

The Force of Mind: Singular Intelligence and SMART Learning

Dr. Max Planck was a theoretical physicist whose work on quantum theory earned the 1918 Nobel Prize. He was one of the greatest scientists of all time, and like Alfred Russel Wallace, saw that underlying all matter and the forces of nature was *"a conscious and intelligent Mind."*

You have been endowed with Singular Intelligence - the Force of Mind. Hopefully this inquiry has opened new avenues of scientific exploration into the wonders of life. However, science alone is insufficient. Objective study demands that you employ all elements of SMART learning. Scientific observation and experimentation must be combined with Mathematical logic, Artistic communication, critical Reasoning, and Traditional wisdom. Only then will you gain a more complete understanding of the many complex issues facing humanity and contribute your Singular Intelligence to the advancement of our world.

We can now return to the question that started this inquiry into Creation, Evolution, and Intelligent Design. **What do you think?**

Notes: